W9-BXM-779

Franklin Watts Inc
387 Park Avenue South
New York
NY 10016

Printed in Belgium

Editor
Ruth Taylor

Picture researcher
Sarah Ridley

Designed by
Sally Boothroyd

Illustrations by
Raymond Turvey
Tony Kenyon

Photographs
Dennis Barnes pages 15(T), 17(T), 19(BL);
Environmental Picture Library 6-7, 18; Eye
Ubiquitous 23(R), 27(B); Chris Fairclough Colour
Library 12, 21(B); Friends of the Earth 5; Hutchison
Library 10(T), 10-11, 23(L); courtesy of London Waste
Regulatory Authority/Geoff Cooper 13(T); Jenny
Matthews/Format 20; Maggie Murray/Format 14,
15(B), 27(T); NASA 13; Joanne O'Brien/Format 9;
Christine Osborne 21(B); courtesy of Pilkington Glass
Museum 10(L); Rex Features Ltd 8, 24; courtesy of
Rockware Glass Ltd 19(BR); courtesy of St Ivel
13(BL); 6.15 Theatre Company 25; Frank Spooner
Pictures 8-9; courtesy of United Glass 17(BL), 17(BR),
18(T), 19(T).

Acknowledgment
The author and publishers thank Tim Wood for help
with research, and Pippa Hyam, Senior Information
Officer, Friends of the Earth UK, for her advice.

Library of Congress Cataloging-in-Publication Data

Condon, Judith
  Recycling glass/Judith Condon
  p.cm.–(Waste control)
  Includes bibliographical references.
  Summary: Discusses the problems caused by the
manufacture and disposal of glass products, and
proposes methods for recycling them to reduce such
threats.
  ISBN 0-531-14077-6
  1. Glass manufacture – Environmental aspects – Juvenile literature.
  2. Glass waste – Recycling – Juvenile literature. [1. Glass
     manufacture – Environmental aspects. 2. Glass waste
     – Recycling.
  3. Recycling (Waste, etc.)] I. Title. II. Series: Waste control.
TD195, G57 C66   1991
666'.1 – dc20
                                89-70742
                                  CIP
                                  AC

WASTE CONTROL

Recycling GLASS

Judith Condon

Franklin Watts

New York/London/Toronto/Sydney

CONTENTS

INTRODUCTION

We live in what has been called a "throw-away" society. Every day, people dispose of newspapers, packaging, tin cans, bottles and plastic containers without a thought. They even dump tires, batteries, refrigerators and cars. This mass of garbage disfigures towns and cities, and when dumped it scars the countryside.

Yet all these commodities have been made from valuable raw materials. Processing or manufacturing them has used costly forms of energy. When they are thrown away, the materials and energy that went into them are simply wasted.

The wastefulness of modern societies has now reached a crisis point, and many people believe that the time has come to change course. We could all help by not consuming more than we really need; and by using materials not just once, but many times over. This helps reduce the amount of raw materials stripped from the earth; it also keeps those materials out of the growing stream of solid waste.

One such material is glass. We think of it as relatively cheap and plentiful – we are familiar with it in our everyday lives. Yet glass is a wonderfully adaptable material with remarkable qualities. Not least of those qualities is that it can be recycled indefinitely. In fact, glass offers something that the fake street seller in Aladdin once promised when he called out "New lamps for old!" This time there is no trick involved. We really can have new glass for old, and in doing so we can help protect our environment, as this book explains.

WHAT A WASTE

The amount of trash thrown out in Britain every day adds up to more than 1.75lb (0.7kg) for every single person, whether adult or child. In Japan the figure is 1.9lb (0.9kg). But the most wasteful country of all is the United States. Americans throw out trash at an average rate of 3.5lb (1.6kg) per person per day, a staggering total of 160 million tons per year – enough to fill a bumper-to-bumper convoy of garbage trucks half way to the moon.

What happens to all this waste? Where does it go to, and how is it disposed of?

In the past these were questions most people never bothered to ask. For centuries garbage from towns and cities was simply dumped onto open ground, often in valleys or worked-out mines and quarries, and then forgotten. More recently, huge and costly incinerators have been built to burn waste. Neither of these methods is ideal. The smoke from incinerators can carry deadly, cancer-causing dioxins and furans into the air. It also leaves behind toxic ash, which in turn has to be disposed of. Landfill dumping, too, creates hazards, as well as being smelly and unsightly. As organic material rots, it produces a gas called methane. Methane contributes to the "greenhouse effect," which threatens disastrous changes to the earth's climate. Where landfill has been covered over and built upon, methane accumulates below ground level, bringing the danger of explosion. Chemicals within rotting waste also filter down to the bottom, forming a deadly cocktail known as leachate. In badly managed sites this seeps into the soil and nearby streams or rivers, poisoning plant, water and animal life.

Is this really a sensible use of the world's resources?

So many garbage trucks, so much pollution, so much waste.

Every item that is recycled is an item kept out of the waste stream. If properly organized, protecting the environment can become a routine part of our everyday lives, right from childhood. Everyone has a part to play.

Understandably, people object to the placement of a landfill near their homes. Angry citizens in Los Angeles, California, have three times blockaded the entrance to Lopez Canyon, in an effort to end dumping there. As the volume of garbage increases, landfill space is running out. Altogether, 28,000 tons of garbage travel American highways every day, and trucks are sent further and further afield in search of places to dump it. The city of New York even loads its waste onto boats, for dumping in other states down the coast. On his first day in office in 1990, newly elected mayor David Dinkins received a message from the city's sanitation commissioner. "Welcome to City Hall," it read. "By the way, you have no place to put the trash."

Now all industrial countries are having to wake up to this same message. The waste crisis threatens our whole environment. If we are to find a solution, recycling will play an important part.

In many countries at present a tenth of the "waste stream" is glass, mainly in the form of containers for food and drink. Yet of all the materials we regularly use and throw away, glass is one of the easiest to recycle. By returning bottles for reuse, and by recycling all glass containers, we can help the environment in several ways. The following chapters explain how.

ALL ABOUT GLASS

People first discovered how to make glass around 4000 BC. Probably they learned from the glass-like material left behind when volcanos erupted or when lightning struck sand. In ancient Egyptian times decorative glass beads were made by melting sand and wood ash. Then a method was developed to make glass containers, by dipping a clay "core" into molten glass. Once the glass had cooled, the clay was chipped out, leaving a bottle or flask suitable for holding wine or oil.

The Romans valued glass, and introduced its use to northern parts of Europe. But it was not until medieval times that glass began to be made on a wider scale. At first, flat glass could only be made in small pieces. Beautiful windows were made from small colored panes joined with strips of lead. They depicted scenes from the Bible to be gazed at in wonder by the congregations of great cathedrals. In time, glass windows, mirrors and chandeliers became a status symbol in the homes of royalty and the very rich.

◁ This glass bottle, decorated with a snake design, has survived from the period of the Roman Empire. It is around 1,700 years old.

Pictured here is the Hall of Mirrors in the Palace of Versailles. The palace was built in the seventeenth century for the French king, Louis XIV. Known as *Le Roi Soleil (The Sun King),* Louis made Versailles the centerpiece of his extravagant reign. Glass mirrors and chandeliers were a symbol of wealth and status.

◁ This beautiful stained glass window is in Bourges Cathedral, France. Similar windows, depicting scenes from the Bible, can be seen in churches throughout Europe. In wartime France some medieval stained glass windows were considered so precious that they were dismantled piece by piece and stored in caves until the war ended and they could be replaced.

American astronaut Edward H. White floats in space, tethered to his *Gemini 4* spacecraft. His protective suit is made of glass fiber.

In the nineteenth century new techniques of production meant that glass could be made in greater variety and volume than ever before. In 1851 Joseph Paxton designed the "Crystal Palace," an enormous building made from 299,655 panes of glass, to house a Great Exhibition celebrating the achievements of the industrial revolution in Britain. The choice of glass was daring, but it proved apt because glass was to play an increasingly important part in the modern world. Glass provides windscreens or windshields for trains, cars and airplanes; lenses for optical use, for spectacles, telescopes and cameras; windows for homes, schools, factories and greenhouses; electric light bulbs; scientific instruments and test tubes; optical fibers for telecommunications; screens for televisions and computers; glass fiber for insulation. And apart from all these, it is still put to the same use as it was thousands of years ago – as an ideal container for food and drink.

At the heart of glass making is a simple process, in which three inorganic substances are mixed together, and heated in a furnace until they fuse. The first is sand, which provides silica. The second is soda ash, which helps the sand melt at a lower temperature than it would otherwise do. The third is limestone, which stabilizes the glass, so that it will not dissolve in water. Various other metal oxides can be added to make glass suitable for particular uses, and also to color it.

Almost all glass manufacture makes use of one other valuable raw material. This material is glass itself: crushed-up, recycled glass.

USE IT AGAIN

One of the most useful characteristics of glass is that it is inert. This means that it does not actively change or combine with other substances. Consequently, glass containers are ideal for holding food and drink. They are not affected by acid. They do not absorb solids or liquids, nor cause them to taste or smell different. And when clear glass is used, for example in fine lead crystal wine-glasses, the sparkling appearance and color of the contents are enhanced. On the other hand, when light would cause the contents to deteriorate over a period of time, colored glass can be used without ill effect.

Because of its natural properties, glass is very easy to sterilize with hot water or steam. This means there is a way to put glass containers to further use without recycling them as cullet. They can simply be emptied, washed and refilled. Perhaps in your own home you have helped to make jam or pickle. If so, you will probably know how jam jars can be used year after year, as fruit and vegetables come into season. The jars are thoroughly washed and dried, then while still warm can be filled with boiling hot preserves. Circles of waxed paper and cellophane, held on with rubber bands, make secure lids, sealing the contents until they are needed. Once emptied, the jars are simply washed and stored ready for the next time.

Every day in Britain 20 million pints of milk are delivered to people's doorsteps in clear glass bottles. At the same time, empty bottles are collected and returned to the dairy. The milk delivery system is an excellent example of how glass containers can be reused.

One of the most famous and successful of all systems to reuse glass containers is the door-to-door milk delivery service in Britain. Roughly three-quarters of all British homes, 15 million in all, have fresh milk delivered each morning. The milk comes in pint-sized clear glass bottles – 20 million every day. The customers are asked to rinse the bottles when empty, and leave them outside the door. More than 34,000 milk deliverers then collect them and take them back to local depots, from where they are returned to the dairy for sterilizing and refilling. It is hard to imagine a more efficient or environmentally friendly system. The customer receives fresh milk without effort. The milk deliverer makes a single journey to deliver and collect the bottles. Often the milkman or woman is a valuable friend to the community, for example noticing when elderly people have not taken in their milk bottles, and reporting to the police that they may have collapsed or been taken ill. Even the milk bottle tops, made of metal foil, can be saved and recycled.

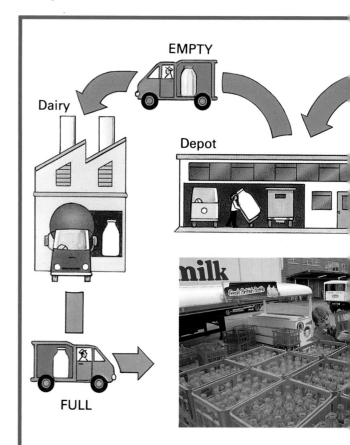

In other countries, it is wine bottles that are reused. In France, bars and restaurants are encouraged to save empty wine bottles for return to wine producers. And in the state of California, a company with the name "Encore!" ("Again!") collects and sterilizes 65,000 cases of empty wine bottles every month before selling them back to west coast wineries.

Another system that encourages the reusing of glass containers is the deposit refund. In the past the makers of beer and soft drinks charged a small deposit on the bottles in which they were sold. When empty bottles were returned to the store, the customer's deposit was repaid. Parents often allowed children to carry the bottles back, and the children kept the deposit. The system made special sense when particular drinks were sold in specially shaped or colored bottles, such as the original Coca Cola bottles. In many countries this system died out during the 1960s, but in recent years it has made a come-back.

This picture comes from France, a country famous for wine production. Returned wine bottles of a standard size and color are being washed for further use.

It is not only milk and wine bottles that can be used again. This photograph shows a collection point for used yoghurt containers in Germany. They, too, are designed to be returned and refilled. Many more products could follow this example.

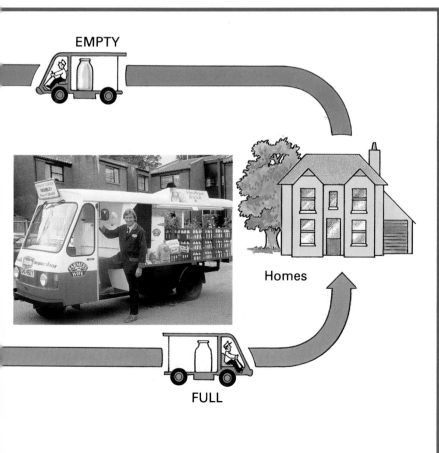

EMPTY

Homes

FULL

COLLECTING WASTE GLASS

Glass manufacturers have traditionally used waste glass as a raw material, because when added to the furnace with the other raw materials it helps the melting process. Waste from factory rejects and scrap has always been a valuable resource. When manufacturers recycle their own waste, they can be sure of using glass made of the correct materials and in the right proportions. This is important if the quality of the end product is to be consistent.

Laminated and other specially treated forms of glass are not suitable for recycling, and at present most makers of flat glass can only recycle their own factory waste. When it comes to glass that has already been used by the customer, by far the easiest to collect and recycle is in the form of bottles and jars. But first they must be color-sorted. The color of glass depends on different ingredients being put into the furnace. If glass of the wrong color is added, the color of the new glass will be spoiled. This is why recycling centers ask for clear glass to be separated from colored, and for brown glass to be kept separate from green. It is also important that metal and plastic lids, corks, and ceramic stoppers are removed, as these "contaminants" make recycling more difficult.

The containers at local collection points come in many shapes and sizes. But all of them require clear glass and colored glass to be separated. The name "bottle bank" is rather misleading. Jars and other glass containers should be taken along, as well as bottles.

Here a container of green glass is lifted and its contents are emptied onto a truck.

There are two main ways of recovering glass containers for use in glass manufacture. The first is to provide collection points, or "bottle collection bins." Members of the general public bring used bottles and jars and place them in large, rigid containers. When these are full, the contents are emptied into trucks, and taken to where they will be crushed and processed. The second way is to provide an organized curbside collection service. Householders are asked to separate glass containers from other waste, and these are collected on a particular day of the week, as part of a municipal collection program or a voluntary plan, or by private haulers.

For the general public, curbside collection is easiest. Some of the most conscientious recyclers are elderly people who may not have a car, and cannot carry their bottles to a central collecting point. For the first system to work well, lots of bottle bins are needed, and they must be close at hand. If people have to drive a long way with their bottles, they are wasting both time and fuel. The ideal collection point is next to a grocery store, or in the supermarket parking lot. Then consumers can dispose of their bottles as part of regular shopping, rather than having to make a separate journey.

Whatever method is used to collect waste glass, safety is an important issue. Broken glass can be dangerous, even deadly. Bottle collection bins have to be emptied regularly and the surrounding areas kept tidy. Workers sorting or handling glass should always wear protective gloves and clothing.

In thousands of homes in the United States, Canada, and elsewhere, glass containers and other recyclable materials are regularly separated out for curbside collection. Most people, if given the opportunity, are happy to make this small effort to help one another and help the environment.

When the green glass arrives at the recycling factory, it is dumped into the appropriate storage bay.

NEW GLASS FROM OLD

The second stage of recycling begins when glass is delivered to the recycling plant. At this stage it is given the name "raw cullet." First it is inspected and weighed, so that it can be paid for. Each load is then stored in a bunker with glass of the same color.

As the raw cullet begins its journey through the crushing process, it moves along a vibrating conveyor belt. Large magnets lift out as much ferrous metal as possible. Inspectors check for foreign objects such as china, brick, or concrete, which are removed by hand. If the glass has come from a bottle

bin, it is usually quite clean. This is much better than glass that has been dumped on the ground and scooped up by a loader-truck, because often stones and dirt are scooped up as well.

Next the glass makes its way into a crushing machine called an impactor. As the glass is crushed, any remaining metal or plastic rings or caps are released from the bottles. The glass emerges from the impactor onto a vibrating screen, from where vacuum machines suck up lightweight materials such as aluminum caps, plastic, or paper. More magnets lift out any remaining ferrous metals. At this stage the crushed glass – now called simply cullet – can be in pieces up to 2in (5cm) long, or in granules, or even in powder form. Once again it is stored in separate bays according to color, before being delivered to the glass factory.

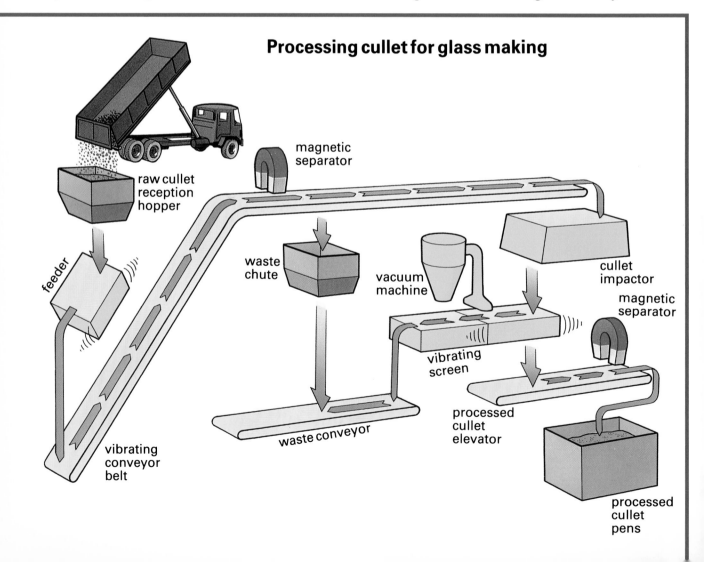

Processing cullet for glass making

raw cullet reception hopper

magnetic separator

feeder

waste chute

vacuum machine

cullet impactor

magnetic separator

vibrating screen

vibrating conveyor belt

waste conveyor

processed cullet elevator

processed cullet pens

This man is unloading waste glass for crushing at a small recycling center in Ohio. Do you think his clothing is suitable to protect him as he works?

In this modern recycling factory the crushed containers are carried along a screen where materials lighter than glass are drawn off by vacuum pipes. At the end of processing, the cullet is dropped into large bays. From here it will be transported to the glassworks and used to make new glass containers.

Here the third stage of recycling is completed. As we saw earlier, cullet of the appropriate color is mixed with other raw materials – called batch materials – and heated with them in the glass furnace. Some glass factories in Europe now use as much as 90 percent cullet in the glass they make. Obviously, in some countries, cullet is produced without the mechanized process described above. In poorer countries, people make a living by collecting waste glass and crushing it in much simpler ways. Sometimes they make glass from 100 percent cullet. The problem with this is that small bubbles of dissolved gases sometimes stay in the glass, giving it a "seedy" appearance. This also weakens the end product – a serious problem in bottles for carbonated drinks under pressure! But skillful glass-makers can overcome such problems. Even small quantities of batch, or of refining agents, produce bigger bubbles which float up to the surface, taking smaller bubbles with them.

So long as care is taken to keep contaminants out, there is no reason why glass used for jars and bottles should not be recycled indefinitely.

WHY GLASS IS BEST

The best and most obvious way to tackle the waste crisis is to waste less in the first place. A major part of household garbage is made up of packaging in one form or another. So, by choosing to buy goods with less packaging, we can reduce the amount of garbage we throw out. But many items do need packaging, so it is also important to consider which kinds of packaging make the most sense.

The qualities of glass containers are easy to list:
1. They do not affect the contents in any way.
2. They do not absorb odors or flavors.
3. They resist chemicals and acid.
4. They can be sealed in various ways, according to their use.
5. They can be transparent, making the contents visible.
6. They can be made in a variety of colors.
7. They can be made relatively cheaply, using mass production methods.
8. They are tough and can withstand constant washing and reuse.
9. They offer the choice of being reused, or recycled as cullet.

No other form of packaging has all of these qualities. Clearly, glass containers are a good practical choice for many kinds of food and drink, and also for pharmaceutical products. In particular, glass is usually preferable to plastic packaging, since most plastic cannot be recycled or reused; and glass bottles are preferable to paper cartons for drinks. Paper treated with wax or plastic to make it resist liquids is not possible to recycle, and paper products contaminated with food cannot be recycled either.

The fact that glass can be reused and endlessly recycled makes it a sound choice in other ways too.

Every ton of glass that is reused or recycled is a ton of material that does not have to be dumped. This saves valuable landfill space, and saves the cost of waste disposal. In addition, every ton of glass that is recycled as cullet saves the use of 1.2 tons of "virgin" raw materials. This is because some of the weight of the other raw materials is lost in the burning process. Saving raw materials means that they do not have to be quarried, and they do not have to be transported long distances from where the quarries are situated. Cullet can be collected from any center of population. These differences mean that fuel energy is saved. The raw materials used in glass-making are not especially expensive or hard to come by. But the use of cullet means that quarrying is kept to a minimum, and this in turn helps to preserve the countryside.

Batch raw materials and cullet enter the furnace together and are melted at a temperature of around 1,500°C (approximately 2,700°F).

Gobs of molten glass drop down and begin to take shape as bottles.

The finished bottles could contain as much as 90 percent recycled glass.

◁ Here sand is being extracted from a quarry in Nottinghamshire, England. Heavy trucks are needed to transport the raw materials to glass factories many miles away. Quarrying scars the countryside, and heavy trucks spoil the quality of life along their route. These activities can be reduced if materials are recycled rather than wasted.

Finally, because cullet melts at a lower temperature than the other materials, recycling helps save energy. The energy used to heat furnaces comes either from the burning of fossil fuels, which contributes to both acid rain and the greenhouse effect, or from nuclear power, which creates radioactive waste. Any reduction in the use of energy is therefore profoundly important to the environment at large. In Britain alone it has been calculated that the use of cullet in glass manufacture has saved the equivalent of 7 million gallons (32 million liters) of oil in a single year. This is a total energy saving of 25 percent.

WHAT ARE THE ALTERNATIVES?

We have seen that glass can be reused; it can be recycled as cullet; it can be wastefully dumped or incinerated. What other alternatives are there?

Many kinds of organic garbage, such as garden waste, kitchen scraps, cotton rags and paper, can be composted to produce a growing medium for plants. Glass, being inorganic, does not rot, and so cannot be used in this way. In fact, glass will survive indefinitely within both open and covered landfill, and can be extracted years later. One company on Long Island, New York, is in the business of mining old landfill sites with tractors to unearth reusable metals and glass for recycling. As the existing landfill is dug out, the bottom of the site can be lined with thick clay and then a plastic liner, to stop leachate seeping into soil and water. The company explains that this is a way of recycling landfill space, and helps make the process economically worthwhile.

In some developing countries, and even in wealthier countries too, poor people pick over garbage dumps looking for materials they can eat or use or sell. Whole families, including children, are sometimes involved. It is obviously an undesirable way to have to survive, and places people in grave danger of disease and injury. It is also a shameful proof of how wasteful societies can be, both of material and of human resources. Clearly, it would be far better if things were managed in such a way that these people had other kinds of work, even if they were essentially handling waste. They could be, for example, collecting reusable bottles, or working in a recycling plant. They could be running a market stall to sell second-hand goods, or craftwork made from recycled materials.

Poor people mend and reuse things that others would throw into the trash. In many parts of the world people subsist mainly on what they can grow or produce themselves. For them it is specially important to preserve and reuse manufactured goods. Such items can only be obtained for cash, and cash is in short supply. So people make lamps and lamp-covers by cutting old bottles with heated wire; children make toys from almost any old container or scrap of material; and in small-scale workshops even glass light bulbs

These people are waiting to sort through garbage at a dump in Mexico. They are in danger of injury and disease, but they are poor and this is the only way they can survive. If things were organized differently, much of this waste could have been collected for recycling before it reached the garbage dump. Then these people would have better work. Wasteful societies waste the lives of people, too.

are mended for further use. Such individual skills as these have been forgotten in modern industrial societies. Yet we can all learn from them. Environmentalists predict that we may need to do so, if we continue selfishly to waste the earth's resources, and those resources start to run out.

When imagination and ingenuity are applied to the subject of recycling, all kinds of ideas emerge. Waste glass can be put to lots of unexpected uses. It makes abrasive paper, "glass-paper," used for smoothing other materials in the workshop. It is used in tiny beads to make reflective paint for roadsigns. It is combined with asphalt to make a road surfacing called "glassphalt," said to have good gripping qualities and durability. It is also used as a strengthening ingredient in compound building materials and bricks. In time, it will probably be possible to recycle more flat glass and other glass products. All these alternatives are preferable to letting a valuable material end up dumped in the trash pile.

A recycled glass factory outside Dacca, Bangladesh.

Tiny beads of glass make this roadsign reflect car headlights so that drivers can see the speed warning at night.

RECYCLING IN ACTION

The whole idea behind recycling is that materials are kept moving in a circle of use, rather than being allowed to drop down into the waste stream. Typically there are at least three stages after materials are first used: they must be recovered; then processed; and then made into new products that people will buy. All of this involves tremendous organization and commitment. It involves individuals, local and national authorities, charities and commercial enterprises. It also requires planning, persuasion, and capital investment. Research needs to be funded, garbage trucks with separate compartments to be bought, and recycling factories to be built.

On the whole, glass manufacturers have been quicker to promote postconsumer recycling than the makers of other everyday materials. They have always been used to recycling their own factory waste, and they also know the economic advantage of using cullet. But the fact that some countries are way ahead of others shows that a lot more could be done. In the Netherlands, 57 percent of the material used in making glass containers is cullet; in Japan 55 percent, and in Switzerland 55 percent. On the other hand, in Britain, glass containers are made from 17 percent recycled glass and in Ireland just 13 percent. Obviously we have plenty to learn from one another when it comes to setting up successful recycling programs.

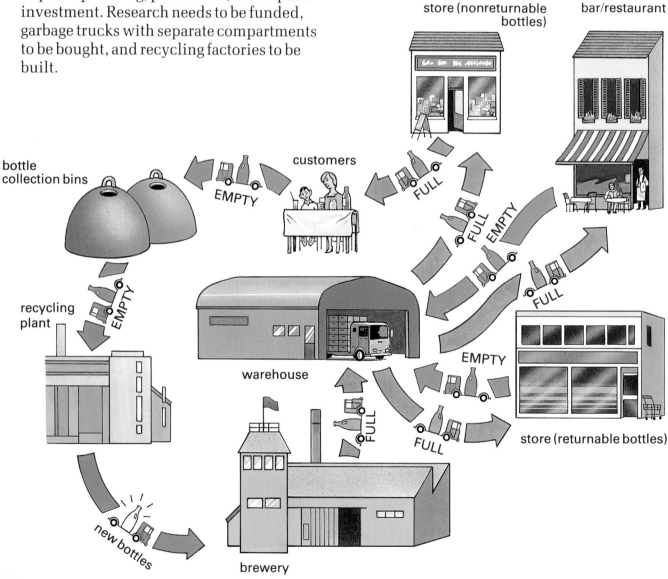

store (nonreturnable bottles)

bar/restaurant

bottle collection bins

customers

EMPTY

FULL

FULL

FULL

EMPTY

recycling plant

EMPTY

warehouse

FULL

EMPTY

FULL

FULL

store (returnable bottles)

new bottles

brewery

In Italy, the bottle banks are bell-shaped. In Switzerland they have been made like a giant dice. But the message is the same all around the world. Recycling makes good sense in any language.

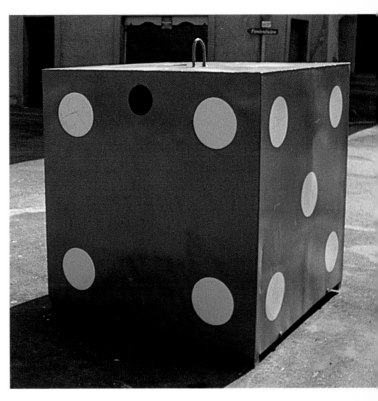

One reason why the Dutch manage to recycle more than half their glass is that they have one bottle bin for every 2,000 people. Britain has had a steadily growing number of bottle collection bins since they were first introduced in 1977, but there is still, on average, only one for every 16,000 people. Yet the British glass industry has the capacity to process more cullet, and actually imports thousands of tons of cullet from Europe every year. A recycling plant at Kelliebank in Scotland annually crushes 8,000,000 bottles and jars into 27,000 tons of cullet for the nearby glassworks at Alloa. And an advanced recycling plant in Harlow, Essex, similarly makes 35,000 tons of cullet from 11,000,000 glass containers.

Britain aims to have one bottle collection bin for every 10,000 people in 1991, but this is still well short of the Dutch average. Another initiative is the UK 2000 project, set up by Friends of the Earth, which has brought together local councils, commercial sponsors and voluntary groups in Sheffield and Cardiff. As part of the project, householders are issued a blue box in which they are asked to put out glass, plastics, and metal cans for collection. Paper is also put out in separate bundles. Most British pubs return beer bottles to the breweries for refilling, but an estimated 400,000 tons of glass bottles are discarded by pubs, restaurants and wine bars every year, so there is still room for improvement. Meanwhile consumers are encouraged to support Britain's unique milk delivery service, instead of buying milk from the supermarket in paper cartons or plastic bottles, neither of which can be recycled.

Industrial nations across the world are trying through many such programs both to reduce waste and to make better long-term use of the earth's resources.

HOW CAN GOVERNMENTS HELP?

Since recycling brings so many economic and environmental benefits, it cannot just be left to individuals, or even to particular industries. At local, national, and international level, governments have an important role to play. They are able to take an overview of what is needed, for example weighing long-term benefits to the community against the costs of setting up a recycling program.

Some governments try to REDUCE waste by placing taxes on products which end up as litter, or by controlling the use of non-recyclable packaging, including plastics. The Danish government passed a law in 1981 which says that all beer, soft drinks, mineral waters and lemonade must only be sold in returnable bottles. Each type of bottle has to be approved by the authorities, and up to now only thirty designs have been allowed. This makes it easier for consumers to return bottles to any shop, and to collect their deposit. As a result of these rules, it is estimated that in Denmark 99 out of every 100 beer and soft drink bottles are returned and refilled. Other European countries which previously imported drinks into Denmark tried to have the law overturned by the European Court. But they lost their case. The court decided that, in this instance, protecting the environment was more important than free trade. Nine American states have passed laws to make deposit programs compulsory, and the West German government also supports the use of refillable bottles.

Other laws are aimed at MANAGING the waste stream. Designers are asked to think about what happens to products after normal use. Manufacturers are required not to mix materials in a way that makes them hard to recycle. In some countries products bear a label saying whether they are made of recycled material. Italy passed a law in 1988 which requires all companies producing or importing packaging to make payments into special associations. Each association is given the task of improving recycling or disposal of its particular material. The association for glass aims to achieve 50 per cent recycling by 1992. If this target is not reached, a "recycling contribution" – in other words a tax – will be placed on glass packaging.

Concern about "green" issues is spreading worldwide. Here Greenpeace supporters from Austria are passing on information to people in Czechoslovakia about a radiation leak which happened there in the past and encouraging them to be more aware of such environmental matters. Active public concern influences government action.

In Britain the 6.15 Theatre Company in association with the British Glass Manufacturers Confederation perform a play called *Bottle Busters* for children aged 7-11. The play is about the discovery, composition and use of glass and about how and why we recycle bottles and jars.

All over the world governments are having to consider how the things we produce, then use, then throw away, harm our environment. Many people care about what are called "green" issues, and in Europe a wide political movement has grown on the basis of such concerns. Governments of different political colors know they will be judged harshly if they do not try to help. They can pass laws and set targets; they can give loans and grants to companies; and they can coordinate partnership projects with businesses and voluntary groups. Government agencies can also set an example themselves by using recycled products, and by reducing waste. The Council of the European Community (ECC) is proposing a wide-ranging set of measures to reduce waste and promote recycling within its member countries. Some of the proposals are about returnable bottles and deposit programs. But such plans take many years to take effect. Governments at a more local level can act faster.

Perhaps one of the most useful ways in which they can help is through education and information. They can sponsor research into the best methods of recycling; they can add waste reduction to the school curriculum, as has been done in Rhode Island and Oregon; they can also fund all the new courses needed at colleges and universities to train the designers and waste managers of the future. Perhaps you yourself might be one.

WHAT YOU CAN DO

Most of us would like to help protect the environment if we knew how. But often we feel things are beyond our control. What can we as individuals do about oil spills on the open sea, or about accidents at nuclear power stations?

When it comes to the problems of waste, however, individual attitudes and actions CAN make a big difference. Each of us can decide to buy things that are good for the environment, and we can refuse to buy things that can cause harm. Recycling can bring people together within the community. It can even be fun.

Here is a list of suggestions about how you can help to save glass. Perhaps you can add other ideas of your own.

★ Don't choose it if you can't reuse it! Whenever possible, buy drinks in returnable bottles, and take back empties to the store they came from.
★ Return glass medicine bottles and jars to the dispensing druggist.
★ Save bottles and jars for use at home. Use them as storage containers.
★ If you have unwanted glass utensils or ornaments, dispose of them at a flea market or garbage sale, not in the garbage.
★ Take surplus nonreturnable glass containers to glass collection centers, or if you have a curbside collection program, put them out on the right day. This includes bottles, and jars for jellies, jams, peanut-butter, baby foods and syrups.
★ Never leave glass litter on the street, or in the countryside. Broken glass is dangerous to people and to wildlife.

Follow the bottle recycling code

- ★ Remove all lids, corks and stoppers.
- ★ Place bottles and jars only in the correct color coded container.
- ★ Take home the box or bag you brought them in. Never leave litter lying about.
- ★ If the collection containers are full, take bottles and jars home rather than dumping them on the ground.

Please save glass . . . . . but don't leave boxes.
Many thanks!

- ★ Find out what recycling plans exist in your community.
- ★ If your supermarket does not allow you to return bottles, write to the manager to ask why not. Explain how beneficial it would be.
- ★ If there is no recycling program where you live, or if the existing bottle collection centers are too far away, you and your friends could write to the local newspaper with your ideas.
- ★ Perhaps your youth club, sports team or school could collect recyclables to raise money. (Aluminum cans may be more suitable than glass.) First make sure that someone is willing to buy them from you. You will also need a safe place to store them.

Take bottles home if the collection container is full.

GLOSSARY

acid rain polluted rain which kills trees and poisons lakes; caused by smoke from the burning of fossil fuels, especially by power stations

asphalt smooth hard surface for roads, made of pitch or tar

batch virgin raw materials to be melted in glass furnace

cancer disease of animals and humans in which cells of the body multiply out of control

capital investment money spent on new factories, machinery, etc

cullet waste glass for use in glass manufacture

dioxin by-product of incineration; one of the most poisonous substances on earth: even minute quantities kill plant and animal life

European Community organization of European countries committed to free trade

ferrous containing iron

fossil fuels substances such as coal and oil, formed from ancient forests and vegetation; a nonrenewable energy source

furan poisonous chemical by-product of incineration

glass fiber finely spun threads of glass, which can be knit, woven, or matted into felt

green name of certain European political parties and popular movements concerned with protecting the environment

greenhouse effect potentially disastrous warming of the earth's climate brought about by gases holding more of the sun's heat in the atmosphere

impactor machine to crush glass by beating or hammering

incinerator furnace where waste is burnt at high temperatures to reduce it by weight and volume

inert without the power to act or move, or to react with other substances

inorganic of mineral origin; a substance that has not grown (see *organic waste* below)

laminated glass glass made in layers for particular uses

landfill place where garbage is dumped on open ground, often in valley or hollow

leachate poisonous mixture of chemicals which seeps from base of garbage dumps

metal oxides various minerals used in melting process, for example to strengthen or alter the appearance of the glass made

methane "marsh gas"; explosive and smelly gas made when organic waste rots

optical fibers very fine glass filaments able to transmit messages by means of light

organic waste waste consisting of materials that once grew naturally in plant or animal form

silica silicon dioxide; basic substance of which sand is made

virgin material raw material not previously processed or used in manufacture

voluntary scheme a plan or program in which people participate out of good will or for a good cause

ADDRESSES AND RESOURCES

The National Soft Drink Association has a variety of pamphlets including *Glass, A Look at Recycling Programs* and *Soft Drink Container Recycling: Questions and Answers.* They also have a quarterly newsletter called *The Soft Drink Recycler.* For more information, write to NSDA, 1101 16th Street NW, Washington, DC 20036.

For information about subscribing to *Scrap: America's Ready Resource,* write to the Institute of Scrap Recycling Industries, Inc., 1627 K Street NW, Washington, DC 20006.

A copy of *Recycling Works!* can be requested from the Office of Solid Waste, United States Environmental Protection Agency, 401 M Street SW, Washington, DC 20460.

An environmental poster and a book called *You Can Do It* can be obtained by writing to The National Wildlife Federation, Earth Day Program, 1400 16th Street NW, Washington, DC 20036-2266. Your name will also be passed along to other environmental and recycling organizations who will send you more information including booklets, buttons and more.

INDEX

PRINTED IN BELGIUM BY

INTERNATIONAL BOOK PRODUCTION